Ms. Cannon lives here on Linden Lane. We discuss many things. Today I admit to a problem.

"I love Gram. But she lives in Japan," I tell Ms. Cannon. "I wish I had a Gram here!"

"I suggest you adopt a Gram." Ms. Cannon says.

I confess I did not think of that.

"What do you expect in a Gram?" she asks.

"Do you happen to have a tablet?" I ask.

I print **Adopt a Gram.**

Sentence one says, "The Gram I will adopt likes kittens and rabbits."

Sentence two says, "She bakes nutmeg muffins."

The last sentence says, "She likes picnics."

This list of stuff is not what I want!

"Can I adopt you?" I ask.

"You bet!" answers Gram Cannon.